For Dei
& gang

Thanks for the decades of help!

Love,
Michele Kish

2036:
THE ALIEN CHRONICLES

MICHELE KISH

2036: The Alien Chronicles
Copyright © 2024 by Michele Kish

All rights reserved. No part of this publication may be reproduced, distributed, or transmitted in any form or by any means, including photocopying, recording, or other electronic or mechanical methods, without the prior written permission of the author, except in the case of brief quotations embodied in critical reviews and certain other non-commercial uses permitted by copyright law.

Tellwell Talent
www.tellwell.ca

ISBN
978-1-77941-582-0 (Hardcover)
978-1-77941-581-3 (Paperback)
978-1-77941-583-7 (eBook)

TABLE OF CONTENTS

2036 The Alien Chronicles: Prologue V

Chapter 1	From Blue To Purple And Back	 1
Chapter 2	Proof Of The First Kind	 5
Chapter 3	Lost Book	 7
Chapter 4	Interplanetary Snail Mail	 9
Chapter 5	Boy Meets Girl	 11
Chapter 6	Why Me?	 13
Chapter 7	Hearts Break All The Time	 16
Chapter 8	The Knight Of Cups	 18
Chapter 9	Love	 23
Chapter 10	Do Or Die	 26
Chapter 11	Now What?	 29
Chapter 12	Digital Dinosaurs Don't Type	 30
Chapter 13	Canine Connections	 32
Chapter 14	Brokenhearted	 37
Chapter 15	A Kaleidoscopic Shift Back To Humanity	 39
Chapter 16	Belief	 42
Chapter 17	Time Is Just A Rubber Band	 45
Chapter 18	Fast Forward	 49
Chapter 19	Lockdown	 51

Chapter 20   Counting Flowers On The Wall 53
Chapter 21   Getting Rebooted 55
Chapter 22   An Alien Presence 57

2036 The Alien Chronicles: Epilogue 59

2036 THE ALIEN CHRONICLES: PROLOGUE

This is a death before life, life after death story. It's too complicated for this human to tell. So, I've decided to let The Alien do it.

It's almost 2024. I'm in a sixty-year-old human body, and time is running out. I have figured out that 2036 will be the approximate year that people here on Earth will get the answer they have been searching for. The answer will be *yes, there is life on other planets*. I'm pretty sure we will hear from the purple people first. After all, they are the closest.

It was 1992 and I was twenty-nine years old at the time. Getting interplanetary snail mail twenty years later was most unexpected. Y'all are really going to like them. I can't wait for scientists with actual degrees to talk to them! My strength is love, not science. Any math other than basic math eludes me.

When I visited the Purple Planet in 1992, we talked about the preservation of their planet. The interplanetary snail mail twenty years later was about our own planet. I keep sensing that time is running out, so I decided to tell this story now.

CHAPTER 1

FROM BLUE TO PURPLE AND BACK

It had to have been a Tuesday because it was mini-meeting night with my meditation friends. It was customary to meditate before meetings. It became the agenda for the evening and all four of us would contribute to it.

That meditation, however, became something else. It was different from the start because I had time. Hours of it. I remember being bored with astral traveling the Earth; I had done it so many times. Saturn was my choice that day. It's my favorite planet, and I was curious what it would look like up close. I felt confident that I could get back in time to go to the meeting.

I grounded myself for longer than usual, given my chosen destination. Grounding happens when you take long, slow breaths in and long slow breaths out, while you bring in the glowing white universal energy into the top of your head. Then, you send the red grounding energy into the Earth through the base of your spine. This exchange presents as an infinity symbol to me, showing me the exchange of love as symbiotic. There are seven protections to put on before you can go anywhere. The first one is a glowing white energy that surrounds you. I personally wear this one all the time because it keeps me safe. The second one is a mirror. Its purpose is to see yourself through other people's eyes. The third protection is a fog on that mirror. It's meant to help you determine if you are being honest with yourself. Four is cool because it is a spinning cloud that is like being in a tornado that flings baggage off when it is worn. Number five is handy if you don't want to get sick. Basically, it's armor. Mine is a Kevlar body suit. Nothing gets through. The Goddess energy of protection six is like lolling in a starlit sky. Number seven is The God's energy. This is real power time. You surround yourself in a golden beam.

Now you can ascend. There are seven planes of consciousness. They coincide with the seven protections. The first three are meant to ensure your focus and intent are strong enough to go on. Ascension four is the willingness to walk through the flames, burning off fear along with physicality. The fifth plane of ascension is where you have teachers to guide you. A quick nod from them and I was on to the sixth plane with the light keepers. They are the ones that will guide me back. The seventh plane of ascension is pretty much your launch pad. The God's energy, that gold beam, becomes your method of transportation.

I remember having strong intent and focus, so getting the go ahead from my higher power was something that I somehow knew would happen. Which was weird because it was the first time that I met my higher power. You know, in person. Face to face.

And then there was Saturn. I was aware of how I looked, dangling there in space. I could see myself clearly enjoying the beauty of Saturn up close. I was in full control of the meditation, then suddenly, I wasn't. Falling asleep during meditation is losing control of it. That's not how I lost control, though.

I felt yanked to the right and cleaved in half at the same time as I passed through a kaleidoscope of intense color. Someone must have pulled the emergency brake immediately after that. I had screeched to a stop at a purple planet. Viewed from a distance, it appeared purple. It's that glorious canopy of theirs.

A window opened and I saw several individuals on a hilltop looking up at me. My first thought was, *who's the alien in this picture?* They are human. At first, I thought they were children because I was so much bigger than them. After their first question, I knew that I was being asked a question that was a very intelligent one, indeed. They looked like regular humans to me. Communication happened through eye contact instead of audible speech. The purple hue from light coming through their purple canopy made all of our skins the same color of purple.

Their question was, how were they to survive the industrial age without ruining that canopy? All I said at first was to never drill into the planet. It was a mistake that we had made. Harness wind and solar power for starters. These don't harm the planet quite like sucking all the lubricant out does, which also makes the planet heat up, plus it ruins the air when used as a fuel. We call it fossil fuel. It's nothing but graveyards that are millions of years old.

At some point I felt like I was snapped like an elastic band. I was returning, rapidly. The only thing I remember is the nod from my higher power as I flew past. Under normal circumstances the return to the grounding position is as methodical as leaving it. I returned so abruptly that my first instinct was to look at the clock. I had become pretty accurate at gauging time during an astral flight. This time, I was off by a couple of hours! *What the hell?* What felt like about an hour had been more like three. I couldn't wait to tell my friends at the meeting.

CHAPTER 2

PROOF OF THE FIRST KIND

When I got to Vida's apartment, Karronne was already there. Apparently, they had been discussing what to do when I arrived. I was expecting to tell them about my meditation, but they had other plans. All they said was, "You took us somewhere," before I was on the floor deep in meditation. They performed the Snap and Tap technique on me.

It goes like this: *Tap, tap, tap* on the solar plexus, while asking a question. *Where did you take us? Snap your fingers for an answer.* I don't know how long this went on, but I woke up with questions of my own. It was exciting to talk about what we all thought had happened. Especially where we went and how we all went together even though we were all in our own homes at the time. We all agreed that somehow, we had all lost control of our meditations, but not by falling asleep.

We turned to our teacher in hopes that she could explain what had happened. Not one of us was prepared for her reaction! She was very angry with all of us—especially me. I was so hurt that I started getting scared. I quit going to classes after that. I never saw any of them again.

My daughter was born in 1991, and my son was on the way somewhere in that time frame. They were born sixteen months apart. Time was something I didn't have anymore.

CHAPTER 3

LOST BOOK

It was in March of 2009. My son has always resonated with all things canine. I swear it's where he gets his sharp instinctiveness. One day he offhandedly said, "Hey Mom, I saw your purple planet on the computer." I confess to complete shock and goosebump overwhelm!

There it was. There was no mistaking it. I couldn't believe we had found it! I had such a science fiction-y belief about where the purple planet was—another dimension. I toyed with the idea of another galaxy as well. To know for sure that it is located on the edge of the next solar system to ours was a relief. That explained the kaleidoscope effect followed by the sudden screech to a stop during my first encounter with the purple planet.

I hadn't thought about that meditation in so long; I barely knew what to do now that we had found the planet. I decided to call Peter Mansbridge, who was the anchor on Global News that covered the story. His advice was to find my journals, especially the one describing the experience. He said that journal is newsworthy, indeed.

It's a plain black composition book marked BOOK TWO. I'm still looking for it to this day. I have found everything else that I have written since 1987. Maybe one day when I'm not looking, I'll find it. At least that's what I keep telling myself. Mostly, I'm pissed off about losing it.

CHAPTER 4

INTERPLANETARY SNAIL MAIL

It was a Tuesday in July of 2012. Once again, I had lots of time to meditate. The Mayan calendar was on my mind a lot. It didn't make sense to me. It became the focus of my meditation that day; I wanted to know what it meant and what I could do about it. I'm uncomfortable with fear. I prefer to face it down. It was fear of the unknown, after all.

I had barely grounded myself and there I was having a conversation with the purple people. Our last conversation had been about their planet, this one was about my own. It was so nice to see them again! They told me that they had followed my advice, and their purple canopy was still intact.

All I wanted to know was, what could I do to change the general consciousness that the world was supposed to end that December.

Study the symbolism of the Mayan calendar, was their answer. I kept hearing the word *Ozabeelay* being chanted, until it was etched in my memory for good. I didn't know what the hell it meant at the time. I guessed it was a greeting, like hello. I know now that it's not a greeting at all. Chanting that word has given me extraordinary powers over the years. I could heal the unhealable while chanting it.

I'll probably never know for sure, but I now believe that Ozabeelay is the purple planet's name. I studied the symbolism in the Mayan calendar, like I was told. The meaning seemed to be a choice: *Go green immediately or die.*

The vision of a healing Earth was an impossibility to me. Wrong again. That vision was back when the planet went on lockdown during the pandemic in 2020. The speed at which Earth healed herself surprised me. Within days the changes were apparent all over the world. Twenty years is no time at all to the purple people! That canopy helps them live longer. I really hate snail mail.

Amazingly, I received an image in late 2014. I can talk to loved ones through memories once they have left this world physically. Memories, especially good ones, are all that we can take with us and leave behind at the same time. I loved my father-in-law so much. That image I received was of him with my friends on that hilltop waving up at me.

It seemed to be a quick stop before carrying on to where he was going.

CHAPTER 5

BOY MEETS GIRL

In 2010, my daughter met her future husband playing a game called *Perfect World*. She was eighteen years old, and my husband and I were never going to let her go to North Carolina to meet this guy by herself. So, plans got made for him to stay with us for a couple of weeks that summer.

No one wanted him to leave. All of us were bawling at the airport. Three weeks later he was back with an open-ended plane ticket. He told us that his father had noticed his inability to concentrate and told him to "come back and get her," if that's what would make him happy. School could wait.

We loaded up the motorhome and embarked on a road trip to meet Grandma in Kamloops. The stars were a hot topic once we got together. Stargazing is loads of fun when you've got someone telling you what the names are. "They aren't the same stars that are visible in North Carolina," he said. I was intrigued.

Astrophysics, we learned, was Raj's hobby. I couldn't resist telling him about the purple planet. He patiently listened, probably wondering about my sanity the whole time. Miss and Raj got married in March of 2012, four months before I got the snail mail.

Eventually, in 2016, he attended Wake University in Raleigh, North Carolina, to study astrophysics. He was happy to tell me about Stephen Hawking and his billionaire friends sending out probes designed to target inhabitable planets. He explained that the scientists' first destination is Saturn. These probes somehow explode, sending smaller probes toward their targets. I must have stopped listening because once I heard "Saturn," I froze.

The gist of the probe technology is that once the probes explode, they travel at three-quarters light speed toward their targets. The returning data travels at half light speed. Bless his mathematical heart, he calculated it up for me, and it would be approximately twenty years before we would hear from my planet. Same as snail mail. Interesting.

So, why 2036? At this point I'm guessing, and I have been wrong before. This body will be elderly by then. What a drag! I'm left waiting. At least by telling you, I'm no longer waiting alone.

So, there it is. The Alien has spoken.

CHAPTER 6

WHY ME?

I was born in the summer of 1963 to ordinary, hard-working parents. I have been told by my older sisters Lee and Lainey that I never cried. They would come to check on me and I would be thrashing about like I was on a mission or something. I was fascinated with the mobile above the crib, and I was trying to reach it. I was born with a bicuspid aortic heart valve. In 1963, they'd say you had a "hole" in your heart. That must have scared the shit out of people.

I'm writing this now, in 2023. Recently, I was scrolling for something to watch on You Tube. The pilot episode of *The Outer Limits* caught my eye. It was made in 1963, and I hadn't seen it before. Strangely, it was about evidence of life on other planets.

My family was busy. I learned the values of having goals and working towards them, from my parents. Some of my earliest memories are empathetic. My thoughts were always positive, and

I enjoyed my own company. I was seven when my baby brother Max was born. To hell with playing with dolls. I had a live one!

My mother taught me how to care for him because it was going to be my job when she was working. My brother was a sweet baby who grew up to be a loving person. Our whole time together was about laughter. Even in adulthood we would start laughing at the sight of each other.

Max died of a heart attack in 2017, at the age of forty-six, one month shy of his forty-seventh birthday. My kid brother had been six years old when our father died of a heart attack; in 1976. I was thirteen. When I was nine, the hole in my heart was renamed a "heart murmur." It must have been getting louder. I could tell it was serious because I could read the body language of the doctor. My life was no fun stuck inside. I had to create fun for my kid brother's sake and my own squirrelly nature. I wasn't allowed to do any sports, especially the endurance kind. So, we found laughter, he, and I.

In 1972, "fresh air" was the go-to answer when doctors had no idea what to tell parents. We were temporarily living in an auto court across the street from our school. We were building a house on the acreage our parents had bought. Our first dog was Snoopy. I loved having pets. We were promised more on the farm.

Grade Five in 1973 was big on cursive writing. Also, creative writing. They called it "long hand," and you were expected to write everything that way. Now, I know how ridiculous this is going to sound in 2023, but we made up contests at recess, during bad weather when we couldn't go outside. One of them was a cursive writing contest. The whole class would be judges if they didn't want to write. It always came down to my friend and writing rival, Mark, and me. He had way more beautiful writing than I did. It forced me to practice more.

In the job I have now, I discovered that I work with a couple of Mark's nephews. It was too late to see Mark again; he had passed away a few years before, unfortunately. Fuck cancer!

Living on the farm was fun because I always wanted to be outside, anyway. There were berries to pick and animals to look after. Plus, my parents were building another house and dividing the property in half. We used to joke about how my father would dream up what we could do the next day, as he slept.

I have had decades to think about why the purple people chose me. The only thing that makes sense to me is that I was the only one dumb enough to be hanging around Saturn at the time. Could be the writing; maybe the fact that I would keep everything that I wrote. Who knows? All I knew for sure, was that was the one day that *I* was the Alien. Whether I already was one, is another question that I have asked myself for decades.

CHAPTER 7

HEARTS BREAK ALL THE TIME

I HAVE LOST MANY loved ones since my father passed. This is why I am convinced that shared memories link us together beyond our physicality. Strangely, I hear from my dad the most often, to this day.

Lucky me! The echocardiogram was invented in time for me to get one on my sixteenth birthday. I wrote my driver's test after the appointment and drove home. I had been told that all the outdoor activities I loved, I shouldn't participate in anymore. Having children would be far too hard on that damaged valve, as well.

I quit school after finishing Grade Ten. I already knew what I wanted to do with my life, anyway. After growing up on a farm, I went off to beauty school. I spent a year taking every course available. My mother cosigned a loan so I could buy a car because I had to drive to the city where the work was.

I remember the day I got the car like it was yesterday. I was leaving the house to go to the first appointment I ever had with Cassandra. I was seventeen, and she was highly recommended for reading tarot cards. Even though a drunk driver clipped the back of my car, I still made it to the appointment. I made an appointment every January after that.

There has always been an underlying urgency propelling me forward. It would leave me feeling like time was running out. I have been going ninety miles an hour with my hair on fire for as long as I can remember! I had made the decision to experience as much as I could. I knew what I wanted to happen. It had to happen before I was thirty.

Eastern practices like meditation and martial arts drew me in for a couple of years. Mostly martial arts. Four trophies later, it was time to move on. It was the early 1980s and I did two important things when I turned nineteen. I wanted to open my own sculptured nail business. So, I did. I wanted to buy my own place to live, not rent it. I did that, too! My father's estate inheritance, which was payable when I turned nineteen, made that possible.

CHAPTER 8

THE KNIGHT OF CUPS

MY BUSINESS WAS booming, and I did anything I wanted because I could afford it. I was still going to Cassandra every January, and 1985 was going to be an exciting year! Lots of travel and more than enough money to make the big purchases that made me happy.

Cassandra read the cards clockwise once they were laid out. *Hellooo, Mr. Nine O'Clock!* The Knight of Cups is a blonde, blue-eyed, tall drink of water. I started asking about time. When can I expect to meet this man? I was twenty-one, and I had accomplished a lot already. Love was on my mind a lot. I got a lecture about time and how it works. Grrr! Stupid time.

Credit cards were a new thing, and I had all of them. That year, I did travel a lot. I bought a 1986 Pontiac Fiero GT, which did make me happy. I was having a very nice life, which I would share with my family. I paid for my family members to share experiences with me. Mostly my mom. My kid brother, too. I loved everyone so much. Money, to me, was something to use to make memories possible. It was somewhere around that time that RRSPs came out. Prior to that, saving money hadn't occurred to me.

There is something strangely magical about having a job that requires holding someone's hand. Sculpturing nails was a career that I had for twenty-five years. I really was quite good at it, and I loved it. In the eighties, nails had to be very long to be considered fashionable. In the nineties, the fashion was sportier, shorter, with a French manicure. My customers were mostly older women. Nailbiters have no age, so them, too. I guaranteed my work. I did not charge for broken nails. The nailbiters didn't bother lying. I fixed them at no charge, anyway.

In 1995 I did my equivalency tests at Douglas College to obtain a Grade Twelve education. Further education could wait. The Knight of Cups hadn't revealed himself yet. I was always super busy around the Christmas holidays. I had close to 150 customers back then. I would blow people's minds recalling their phone numbers. Remembering stuff is much easier if you write it down. I made a game out of it. I've been in an entirely different career for the last twenty years, and I still use that game. The only difference is, I take a picture of the numbers with my eyes to remember them.

I loved my customers, and they proved their love of me in return, by spoiling me at Christmas. I would have a ridiculous amount of presents under my tree every year. I brought a box to work every day in December, just to make taking all the presents home easier. I miss every single one of them to this day.

I really liked the band Shriekback. I had bought a ticket to their show on the seventeenth of November 1985, at the Commodore

Ballroom. I will never forget the powers that be telling me that the man of my dreams would be there, as I was getting ready to go. I made sure that I looked perfect. I was disappointed when I didn't just recognize him. Pissed me off! Maybe he simply wasn't there. Whatever.

I took skiing lessons on the local hill that winter. Although I recognized how it could be loads of fun if you knew how, I wasn't willing to go through the pain involved in learning. Too many faceplants, not enough control.

In January 1986, I went to see Cassandra. Where the hell was that Knight of Cups, anyway? *Well, well, well. There you are, Mr. Nine O'Clock.* I got that lecture from Cassandra again, about time. I was sick of waiting.

I mentioned how buying a new car made me happy, right? There was a nightclub over an hour's drive away that I had never been to before. So, off I went, that Friday night, driving stick in stilettos. Was I ever overdressed! I was trying too hard with my appearance; I had been for a year, already. I never got asked to dance. Everyone, women included, couldn't seem to take their eyes off me. *Screw it! Time to go.* I had to walk to the back to use the washroom. I kept my eyes on the floor because the worn-out carpet was a trip hazard, and I was in five-inch heels.

I took maybe six steps after leaving the ladies room before the Knight of Cups reached out and pulled me into his arms and asked me to dance. It took me five months to realize who he was! That night, I gave him my business card. Landlines being what they were back then, my home phone number wasn't on it. The following day was a Saturday, and I worked on Saturdays, but I took Sundays off.

My Knight of Cups had a name: Les. He called and mentioned his plans. My uncle Tom had retired from being a longshore worker, and I was going to his retirement party. I left the party to meet up at the bar Les was going to be at. I tend to be early for everything. I was starting to get approached by guys, so I decided to take off and go back to my uncle's party … I had blown my chance to give Les my home phone number. Crap!

Luckily, he called the shop on Monday and asked if I had been at the bar, because he had looked around for me. I lied. I was mortified by how quickly I had been captivated by this guy. I didn't want to admit that I was so anxious to see him again, that I was too early! I just said that I hadn't been able to make it. *Ugh! What a liar!* I got exposed after we had been married for ten years, though. No one cared.

It was 1986 and Vancouver was having an exhibition that year. Expo 86 began in March. Almost all my dates with Les were spent there. I was offered season tickets by one of my customers. Her husband had bought them for his staff, and they had extras. By then, Les and I were pretty much living together. That first Monday, he came to my place with a bottle of twenty-five-year-old Grand Marnier; I handed him the keys to my car the next morning.

Transit was taking too long. Plus, Les told me he would take care of the maintenance of my car. I had a partner: someone there, wanting to make my life easier. That is winning the lottery! Les told me that the first night he saw me, his mind demanded that he reach out and grab me or he would regret it for the rest of his life. We felt married from day one. Both of us were choked that we hadn't met at the Shriekback concert, though. The time wasn't right, or the stars weren't aligned, I guess.

CHAPTER 9

LOVE

That September, my family had agreed to sell the fourplex and buy a tri-plex closer to my business. I converted a room into my new shop. Les moved in. That dude could fix anything! My mother was having trouble with my kid brother not wanting to finish school, and I had a customer to fix that. She was a teacher in an alternate school. I had enrolled Max to start in September, which is why my family had to move close to my business. Plus, Les was already involved in my life, and intent on protecting me.

I wasn't used to someone loving me and taking part in the pressure that my family put on me. I was so friggin' jealous all the time! He was so perfect; I couldn't stand it. My heart murmur was getting louder so I was prescribed blood pressure medication at twenty-two years of age. Les called them my "heart pills."

Full of surprises, Les was a ski technician. He was surprised that I had taken lessons just before we'd met. I didn't come off as the outdoorsy type, obviously. My heart has dictated all my activities my whole life. Picking my ass up off the ski hill seemed too hard!

One thing my heart did dictate was the value of giving it a workout. I aerobicized. It was the eighties, for heaven's sake! The tunes were stellar. Madonna's first album was a go to. It was a long-playing vinyl album that I began using to exercise at home with. Micheal Jackson, too. Les said he would take care of me and teach me how to ski. He convinced me through action. He was a ski technician. I had equipment that was made for me. Les watched as I progressed and cranked my bindings up to match. I had Mondays off back then; one Monday I had decided to go to Whistler by myself because everyone else was working.

I learned to let my skis run that day. I was so fuckin' hooked. Whistler was where it was at, alright. I felt grounded. I had to agree with Les; we had to figure out how to live there. It was our destiny.

Time was running out, though. The man that I was with was the right one … so, why was I pushing him away? I was such a bitch, and I hated myself for it, which just made it worse. I knew the problem was mine, but I was reluctant to own it. Luckily, I had customers for that. I have told you about the meditation teacher; I had been introduced to her through a couple of my customers. I had to fix whatever the hell was wrong with me if I wanted Les to stay!

Love is the only true power, was the first thing I learned. Love of the unconditional kind, however, is inherent to canines, not humans. I have a whole other chapter on that, though.

Learning to meditate saved my life as well as guided it. It was 1987 when I began writing *The Alien Chronicles*. I had started up with it in the first place because I wanted to deal with my own stuff instead of expecting my partner to. I've come to realize that this is why marriages don't work out sometimes. If one of you has a suitcase of baggage, it won't do to have a partner with a U haul of their own!

CHAPTER 10

DO OR DIE

I DID ABSOLUTELY EVERYTHING that I was told *not* to do, and it didn't kill me. I trained to climb the west Lion's Peak in Lion's Bay by walking uphill for months, and that didn't kill me. There was one last accomplishment before I turned thirty. Kids: the ultimate no-no, from a health perspective. I was in good shape, and the medication I was prescribed seemed to be working. Les and I had to get married for two reasons. First, I had to trade stronger heart medication for the birth control pill. The Pill had to go because it was too hard on my heart. Secondly, my parents-in-law were religious, and we didn't want to hurt them. So, we got married.

A very, very strange experience happened within a month of Les and I getting married. I mean, strange as in a noun, person, place, or thing: Les Paul Strange. This dude is memorable because of that encounter. Les and I were skiing, like we did every winter. We were riding in the gondola when I noticed a unique watch that the two others in that gondola were wearing. These watches were custom made by the Exxon company, as gifts to anyone who showed up to clean up the mess the *Exxon Valdez* made on March 24 1989 when it ran aground. It had to have been worth a fortune. I looked up into the eyes of the dude sitting beside me. *Wait, what? Did my man just get twinned??*

These passengers were fishermen from Michigan that just happened to be close by when the *Valdez* went down. Apparently, Exxon was grateful for their help; skiing at Whistler was an extra. We were four people in a gondola. It was not lost on any of us that my Les and Les Paul Strange looked like twins. They even sounded alike.

It was impossible to say goodbye, so we invited the fishermen to crash at our place. I was having trouble staying in the hot tub, and that's when my husband's twin told me that I was pregnant … and that I was having a girl. *Yeah, right!*

I started puking a couple of days later. I was pregnant, all right. And we did have a girl. My doctor took me off my heart medications right away. My poor heart! I relied heavily on meditation to bring my blood pressure down. In utero, my daughter was in a posterior position, which is not ideal for delivery. I had to be admitted to the hospital a month before she was born. Once again, I wasn't allowed to do anything.

That bicuspid valve was preventing both of us, mother, and child, from getting adequate oxygen. We were lethargic. Obviously, Miss wasn't going to do what anyone else wanted her to do, so she was born by Caesarean section. My blood pressure was 210 over

170 when I woke up. We were told that a possible stroke was the reason for the advice we were given, to not have any more kids.

I don't know about anyone else, but if I get something that is just so great, I want another one. I had that thought one day … and I was puking my face off the next. A boy revealed himself in the ultrasound. One of each, thank God. Les and I knew that five kids were out of the question, anyway. I was a couple of months shy of thirty when Matt was born.

CHAPTER 11

NOW WHAT?

While writing this story, I have also been reading my journals. I can't believe how many times I have tried to write this story. New developments, new beginnings, over and over. Most of them were too early. I know that at this moment a probe is headed to the purple planet. I have always maintained that I have been there and that there seems to be a connection having to do with planet preservation. And there are people.

I will be left wondering, like I always have, until the next development. The next chapter in my story will be that final proof of life on another planet. I can pass the baton to those that can put their education to the ultimate use.

I have believed for a long time that I get to go to the purple planet, after leaving this one. That is the final chapter in this story. I have seen myself on that hilltop with my friends, waving up at you. Now wouldn't that be the limit! Too bad I won't be here to see it.

CHAPTER 12

DIGITAL DINOSAURS DON'T TYPE

THE LAWYER THAT Les and I use has a degree in literature, as well as law. He insisted that I learn to type. I dug up an antiquated tablet that I never used and wrote this story for the first time, without a pen in my hand. I was getting all kinds of good advice. Speaking into the microphone for talk to text, I thought that was the ticket. Until I noticed it wasn't typing what I was saying! My daughter has "kaleidoscope" eyes, not "collateral" eyes. *Grrr.* The one thing that I have found interesting is that I can somehow summon the purple people's help. Only since working on this book, though. I have chanted "Ozabeelay" before, and it is effective. I can heal, and I can send energy chanting it.

Off the top of my head—I don't know the exact date—but Jill Biden was in Seattle that day. Lee, my sister, got caught in the traffic after leaving my place. It was a Friday. I was worried because Lee is seventy-six-years old. So, I kept calling her place and talking to her husband. That's how I found out the traffic was bad and why.

I called three times, and my brother-in-law sounded the same each time. My sister had to take him to the hospital as soon as she got home because he had a heart attack, apparently. I couldn't compute. How come I didn't notice? I texted my sister as they waited around at the hospital.

At some point between texts, I decided to send love and the purest energy I could generate. I positioned myself facing in the general direction of Tacoma, Washington, where they lived. I began chanting *Ozabeelay* and powering up. I usually send energy in a straight line. I kept doing this in between updates.

Then, the purple people showed up and taught me how to throw a curve ball. I never saw the energy bend like that before. I instinctively knew that the hospital was targeted. That's when they helped me to really pour on energy.

The next day, my sister called and made me explain what happened. She told me that she was thankful that I kept calling. She believes that I wasn't worried about her getting home safely. I was somehow looking after her man until she got there. I hope so. They are in their late seventies. I love them both with all my heart.

I don't suspect an end to this "purple effect," in my lifetime. It's kind of neat that they come to call on me, though. I am going to figure out if group meditation might bring them around.

CHAPTER 13

CANINE CONNECTIONS

Dogs have that one capability that eludes humans: unconditional love. Every single dog has a character of his own. Snoopy was the first pooch that we got once we were living on the farm. He was a high energy dog that loved to scope out the ladies and often ended up in the pound.

Back then, hunting was allowed in the area our farm was in. We would find empty shotgun shells all over the place when we rode our horses. Snoop was shot in the chest, once. He must have just walked off one day because I don't remember what happened to him.

Then, we had Sport. He became my kid brother's shadow. Farm dogs are far too stinky to let them in the house. When my dad died, Sport would lay by the car waiting for him every day. I let him come into the house one day when nobody else was home. He figured it out. I can't recall what happened to him.

Tuxedo Max, Isabelle, and Rokko were a pack. Les and I adopted Max in 1998; he was a very smart Dobie/Lab cross. He belonged to a friend who had died in a work-related accident. We couldn't change his name because he already answered to it. My kid brother didn't care, so why should we?

Max saved my life one day, when I forgot that I had put some bread in the toaster. I was upstairs, and he kept on me to come downstairs. I could see smoke in the air, and the flames were catching the cupboard on fire, already. After putting out the fire, I searched for a big piece of meat in the fridge. The meat went down the hatch before the smoke alarm went off! Max wasn't fixed, so he would disappear a lot. We tied him up, and the females would come over and visit him, instead.

Then, we adopted Isabelle off the reservation in Pemberton, in 2001. She had coyote in her, and she was very pretty. She had seven pups in 2002. The third puppy was born unresponsive, and Isabelle didn't bother with it. Les replaced the dead puppy with a stuffed coyote. It confused the hell out of Izzy, so he flung the stuffed toy down the hallway. She calmly went and got it and put it with her puppies. She treated that toy like one of her pups. The pick of the litter, in our eyes, was Rokko. The other pups were pounced upon. Our Tuxedo Max was smart, and my customers were lining up for a pup. We kept Rokko because they had all become a dog pack by then.

Max never touched the garbage after being pile driven into it once as punishment. We were stuck with Isabelle and Rokko tag teaming the garbage, so it became their turn. Isabelle wasn't having it, though. She struggled so much that the second she landed in the bag, she blasted out the side of it like a rocket. The hilarity of coffee grounds flying in her wake made it hard to get back to the task at hand. Rokko was perfectly happy in the garbage bag. At least he learned to leave it alone.

Isabelle wouldn't stay out of the garbage, though. She died when she was four. The kids started working in 2012 and we had to rent a house closer to the city where we all worked. Max was seventeen years old when he disappeared one day. Each one of us did our share of crying, for five days. Then, a police constable knocked on the door one day, and Max came trotting in. He explained that Max had been walking down the middle of the street like he was on a mission, so he'd followed him.

Meanwhile, Rokko was inside giving Max shit. An elderly dog always looks too skinny. One look at Rokko and the officer's mind was eased. During a road trip the following year, Max disappeared. Rokko was always able to bring him home. We stayed for four more days, hoping Rokko would pick up his scent. The dog shelters didn't pick him up, either. Getting old sucks. I know now that Max didn't like being in his old body. He didn't want to be found.

Rokko would have been ten years old before he got to be an only pet. He was awesome. I would walk him every day. He was obedient off leash. He liked fetching balls like his dad. Losing Rokko was hard, and no one thought a dog existed that could ever replace him. Then, came Zephyr. But first, Rokko's finest hour. He was the only pooch that I performed a healing on. It worked, too! We think that he was hit by a car because he limped occasionally. Anyway, in 2012 we still had tenants in our Whistler house, and we were all in the city working a lot.

So, Miss and Matt took the bus up one weekend and told us that there was a family with a nasty pit bull camped out in the living room. Matt had seen them before, so he convinced Miss to call us on her cell. The landline couldn't be used because it was in the living room. Plus, the pit bull lost its mind when anyone in the house moved.

After work, we packed to leave instead of waiting until the next day. It's a two-hour drive so three in the morning was our estimated arrival. Miss and Matt waited patiently in the dark for the fireworks. Rokko was all hackles when he realized what was going on inside his house. We could hear the pit bull whining on the other side of the door. Rokko couldn't wait for that door to open, because once it did, he charged up the stairs after the pit bull.

Our dogs were never allowed on the furniture. Rokko was choked; the pit bull was up on the pull-out couch trying to hide behind its owner. While the dude was wearing his dog as a hat, we realized the whole house was awake. I remember saying, "Whoever you are, get the fuck out!" His woman yelled, "Who do you think you are? We have been staying here every weekend! You have no right," and so on.

No one attempted to restrain Rokko, so the pit bull landed in their vehicle quickly. I was enjoying myself. Les was roaring at the tenant that let them stay without our permission, while I dropped my voice and said, "We are the owners of this house, and you have stayed here every weekend? Is that so? Get the fuck out!"

The tenant that let them stay left the next day. The other tenants in the house were using the washroom one after another, and I asked each one why they didn't say anything to us. No one wanted to be a rat. Suddenly, snacks were flying all over the place. Everyone wanted to show their appreciation to our Rokko. After I wrote this, I realized that I had to remember it first. I love memories.

CHAPTER 14

BROKENHEARTED

I FIND THE TIMELINE between Rokko and Zephyr worth mentioning. When my kid brother Max passed away in late 2017, we had to road trip to the funeral. We noticed something was wrong with Rokko's leg almost right away when we got back.

He underwent surgery but the problem came back. Matt bought a house and Rokko didn't have stairs to climb anymore. That was March of 2018. Zephyr would have been a year old by then. He was born in Modesto, California. He's a pure-bred Siberian husky, the poor bugger. Everyone wanted a dog like the ones in Game of Thrones. I hadn't watched a single episode of it.

We all said goodbye to Rokko as a family, in July of 2018. His ashes are with his mom's. I had never been more depressed in my life. It kept getting worse, and I noticed that we were *all* depressed. We had no reason for it. We bought a boat, and we started having fun again. Why weren't we happy? We were pooch-less and didn't know that being without a canine was the actual reason we felt depressed. Something was missing.

During that time, I started dreaming about wolves a lot. We moved in with Matt, and we were kept busy tackling the jungle in the backyard. There are three massive cedar trees, two of which are in the back. Being among trees is grounding; they are just as powerful as a mountain, as far as anchoring goes. 2019 came around, and I started to think about getting another pooch. I was the only one, though. Everyone else wanted nothing to do with the responsibility. I could sense that someone was looking for me at the same time I was searching for them. His name was Kuma, Japanese for "bear."

CHAPTER 15

A KALEIDOSCOPIC SHIFT BACK TO HUMANITY

Teaching Miss to meditate began early and came easy to her. Matt had a bandwagon full of celestial guardians that made sure that he understood, even though he chose not to participate. I loved how Miss would run with it, though.

One time, she was practicing invisibility in the middle of a snowstorm. She sensed a presence and opened her eyes, and a coyote was beside her, looking in the same direction that she was. Magical! Either the coyote didn't see her or was wondering what she was looking at. Whichever it was, I'm sticking to magic. Invisibility isn't an easy focus to hold once you have a fear of discovery.

Get this: I had put a message on a massive white shirt when I was pregnant with her; you know, because I was whale? It read, "Master Under Construction."

During meditation, I had the extra focus of a marshmallow protecting her. Too much information may pose a danger. I underestimated Kaleidoscope Eyes, though. *Ay caramba!*

So sneaky! I was resting up for a graveyard shift one evening when the kids were still in high school. I got a panicked call from the mother of Miss's friend, asking if I knew where her daughter was … because she was not at my house with Miss, like she said she would be.

Les was furious, and so was I. I called the house, and Matt said they weren't there. He didn't want to fink on his sister, but he told me that they were in the bathroom for a million years putting on makeup. Then, they took off. *Hmmm.* Being that it was St. Patrick's Day, I knew immediately: Packed bar. Stealth. Too easy.

I told Les that I knew where they were; I just needed a couple of minutes to find proof. I didn't need to ascend. I rode the gold beam. It was local, like 100 miles from my location, not very far at all. Of course, I didn't really know how long it would take; I was rusty, after all. *Woohoo!* I was back within five minutes, telling Les that I saw the little shits through the window of the pub.

I immediately got on the phone to the girl's mother. She knew the bartenders and wondered why wouldn't they tell her? "They are hiding behind a pillar," I said, "and doing a good job of it." She called the pub before going down there, and no one had seen them. She called me back, and I told her exactly where they were. I saw them through the window. "They are standing behind the pillar that is closest to the door," I said.

It wasn't long before I was called back. "They were exactly where you said they would be," I was told. "How could you have possibly known that?" "It's a long story," I answered. "I'll tell it to you one day." Looks like that day is today.

It was time for me to go to work by then, and I shook it off and focused on Les's face. He had witnessed everything, and the size of his eyes told it all.

Once I was face to face with those nonchalant, kaleidoscope eyes, I asked them what happened. The bartender spotted them, they said. Are you sure about that? Was her friend's mother strangely silent?

Those mesmerizing eyes became uncomfortable, suddenly. I described how I had to astral travel, looking for her, instead of resting for work. The doubt vanished from my daughter's eyes when I described in detail seeing the two of them drinking light-colored beer. The kicker was when I said that the two dudes behind them, checking out their asses, were drinking dark-colored beer. Busted!

CHAPTER 16

BELIEF

IN 1986, VANCOUVER hosted an exposition. In 2010, it was the winter Olympics. The slogan was "BELIEVE." That slogan was responsible for every medal Canada's athletes won. It was genius! Why do I think this? There is a saying about believing it when you see it. This slogan, however, puts focus and intent into the equation: I WILL SEE IT WHEN I BELIEVE IT. Forechecking, if you prefer a hockey analogy. Our Sydney Crosby was relentless, remember? Forgive me if you weren't born yet; I keep forgetting. Canada winning the gold medal in men's hockey was unforgettable in 2010!

I will always maintain, "you'll see it when you believe it." It was trippy having all this going on in our backyard, while we were watching it on television. The crowd's roar from outside was always louder than the TV.

By March, Miss was playing *Perfect World* with Raj. Then, within a month, they were just talking, mostly. They both admitted they were playing the game just to talk to each other. One day, I happened to be in the kitchen, and Raj asked Miss who the old lady in the background was. Miss burst out laughing, and I insisted on being included. She told him that the "old lady" was her mom.

Fine! I had stumbled into bad lighting. My turn! I calmly walked into view, flipped the double-whammy, and introduced myself, smile included. I was enjoying myself; he was uncomfortable. Being a parent means that I have earned the right to be this way.

So, Raj lives in North Carolina. They were both born in 1991, so Miss was still eighteen-years old. After about a year of flying back and forth, they decided to get married.

They gave us ten days to whip up a wedding. The vision of my daughter with sneakers and a wedding dress on, running all over the gym we had rented on the fly, was nice to watch. Happiness.

Then, they had to wait for a year for an immigration appointment in Montreal. So, Miss and I flew to Montreal where the beggars on the street speak perfect English, and everyone else tells you they don't speak English. In English.

The hostel was awesome, though, and the young people staying there were very friendly. It was decorated with Hollywood features, and it was so old that the wiring seemed to be a series of extension cords. If you adjusted your position, you would lose your internet connection. The first night, we listened to Jean-Michel Jarre's album, *Oxygène*. We were imagining he was performing on the roof, like in his video.

Miss was so nervous about the inquisition she was about to subject herself to. Off she went looking very chic, which was tough to do because she had to walk through knee-deep snow to get to the appointment. I knew that body language could betray you, so I wasn't nervous at all. I knew their love was real.

I had us all packed up to go to the hotel by the airport when she returned. Once we got there, and the novelty of getting beer out of a vending machine wore off, Miss told me everything. The interviewer took one look at her and asked her how she had met Raj. Miss found herself just telling her story, and not needing to prove anything at all.

She thought it was weird until I asked her if the gal was wearing a wedding ring. The light bulb went off in her head when she realized the love that she had found, had shone through at the interview.

We flew back to Vancouver, and I became daughterless.

CHAPTER 17
TIME IS JUST A RUBBER BAND

THE KIND GAL that was assigned to edit this book told me to stay chronological and not to repeat myself. "Restructure it," said she. I deleted huge chunks of text while my husband laughed like crazy. He tells me all the time that I jump back and forth when I am telling a story. It makes me hard to follow. Damn it! I'm going to do it again! Ugh!

When I was a teenager, I struggled with what to believe. I even attempted to read *Dianetics* by that Scientology dude, L. Ron Hubbard. I admit to sensing narcissism, and I decided to put it down. I always seemed to end up reading what I needed to, though. The Bible took me five months to read. I read it because I didn't want anyone else telling me what was in it. I was surprised to find my last name; it's in the Old Testament. The New Testament is way more fun to read, though. Jesus must have had a pretty good sense of humor. I used to waltz the Jehovah's Witnesses in for tea every Wednesday to study it.

I had a blood transfusion as a baby, and I was curious to read exactly where it was in the Scriptures that gave them their belief to *not* believe in blood transfusions. When I mentioned it to Les, he said that if the words are in the Old Testament, they couldn't have possibly known about blood types! I hadn't thought of that.

Another book that landed in my hands was *The Power of Thoughts*. It's an eastern Bible. Prayer and meditation, especially in a group, is powerful. In the classes that I took, we were guided through meditations. Finding them word-for-word in bibles, years later, pissed me off. I have gotten over it, though.

It's 2024, and I believe in myself. Belief takes time. Ideas can sit on the back burner for decades. Suddenly, if you don't act upon them, they boil over on the high burner, in your face—like the pressure that I felt to write this book.

I write. I can't help myself. On paper, preferably. I like pens so much that I got lots for Christmas. 2023 has forced me to shed my endless whining about how technology has seemed to evolve without me, while at the same time forcing me to reflect on many things that I had forgotten about. These memories have been on the front burner the most.

A few years ago, my Miss grew the huevos to get a DNA test. I was happy because many lifelong questions of mine got answered. There was an ancestral physique thing in there that answered the question why I could always win the teddy bear at the PNE

(Pacific National Exhibition in Vancouver). Muscle, which weighs more than fat, got proven every time I stepped on the scale.

Having to stay in shape, once it became known that exercising your heart was important even though it raised your blood pressure briefly, was a green light for me. Aerobicizing! It was the 80's, after all, and the tunes were stellar. Music and dance are ancient and tribal, and felt to the bone. Canada is full of First Nation folk that have lore passed down from generation to generation. The fact that British Columbia has finally recognized that ceremonial burnings have an ancient purpose in this province, made my heart sing out loud. Controlling the number of forest fires we have every year, sounds good to me.

I have had odd experiences over the years that I would like to share. The year had to have been around 2009. I was in the frozen food section one day at the supermarket. I was on one side of the aisle and a young mother holding a baby was on the other. I glanced at the baby who was staring at me like I was a television. I found myself walking over and saying, "Ah, there you are."

The mom didn't notice at first that I was talking to her baby. It was when the child tried to escape her mother's arms and land in mine that I got her attention. I have never encountered a baby that isn't shy with strangers. Suddenly, I was mad at myself because a child thinking strangers are okay isn't a good thing. We knew each other though. I forced myself to apologize and pretend to shop. I completely forgot what I was shopping for. It was hard to take my eyes off the child. She was trying to keep me in sight all the way down the aisle and around the corner.

Another time, I was working with my longshore sister, Linda, while she was pregnant. In fact, we worked together during both of her pregnancies. I looked at her one day and saw a cherub surrounded by hearts. Her daughter was born on Valentine's Day. Then, her second one caught my attention one day. The little monkey wore an impish expression as he demonstrated his superior climbing skills and dangling abilities. I was not surprised to find out that nothing was safe from his climbing.

A time warp seemed to happen with another work sister named Karla. We were partnered up in two kiosks, working with truck drivers that were taking containers out the gate. It wasn't that busy, and we were talking about meditation a lot. I offered to do a guided one over the lunch break with her.

We pulled down the shades and somehow lost track of time. It had been a long time since I had meditated with anyone; more places to go kept popping up. When we came back and opened our eyes, we were surrounded by truck drivers that were unwilling to disturb us. Neither of us heard their trucks pull up. Technically, we were twenty minutes late for work! It was hard to believe we didn't hear them pull up, though.

CHAPTER 18

FAST FORWARD

Remember "Game of Thrones"? Huskies became popular all of a sudden. They are hard to train because they are perpetual puppies. I don't mean training them to do their business outside, I mean how annoying they can be. I was having wolf dreams a lot in 2019. I wanted one, and I couldn't stop thinking about it.

I began searching for one who was available for adoption. I found out about a wolf dog named Kuma. I went to meet him, and I was smitten. I felt sorry for him because he came from California, and he is a winter dog. He must have been expensive for his original owners because he was a purebred Siberian husky, microchipped and all. Escaping was his specialty, I was told.

My birthday was the next day, and I was never denied what I wanted on my birthday. I wanted roast beef for dinner ... and a dog. Kuma was two when he became ours, and he was a skinny puppy. He had been running feral in California when he was picked up by the pound.

There was no way that I would put him in a kennel, even though I had been warned that letting a husky roam the house at night is not a good idea. I didn't want to make him wear a collar at home, either. After about two weeks his tail suddenly had a life of its own. We renamed him Zephyr because of the American skateboarding team, the Z-Boys.

He had been trained to jump up for a hug. That had to go because our nieces were half his size. I hated having to knee him in the chest. The last time he jumped at me, I punched him in the face without thinking. That really broke my heart.

Then, my heart broke for real.

CHAPTER 19
LOCKDOWN

I was walking across the floor one day when it occurred to me that I was going to faint. I managed to lock eyes with Les, who sprang out of the chair like a tiger, and caught me before I hit the floor. An echocardiogram revealed an image of my aortic bicuspid heart valve slamming together like two swollen tongues. It was December 2019, and I had made the valve last for over fifty-six years.

Going to work wasn't a good idea, Les insisted. Longshore workers that are in danger of passing out on the job shouldn't be there. It's already a dangerous job that requires being aware of all the heavy machinery that's constantly moving around you.

The first thing that I noticed: my doctor became all singsong and nonchalant as she described what had to happen. I know that she was trying not to scare me, but fear snuck in, anyway. I had to meditate to cope while I listened to her say that a stent had to be put in, first. Then, three months later after that was healed,

I would get the open-heart surgery to replace the defective valve with one from a cow or a piggy. They didn't seem to want to say which.

Then, the COVID pandemic hit.

I started churning out homemade masks the size of diapers in between gawking out the window every two seconds like a holed-up gangster. I worried about the waiting time for my stent procedure because the hospitals were being overrun in no time.

CHAPTER 20
COUNTING FLOWERS ON THE WALL

THE COVID TEST wasn't even invented by the time I was called to get the stent put in. My care card was accepted via a ten-foot pole from the administration nurse in a hazmat suit hunching behind what appeared to be a wall of riot gear. The nurses and doctors were talking amongst themselves. They were going on about lockdown, and each one of them was trying to make others laugh.

I had been having trouble getting that counting flowers on the wall song out of my head. It was from the movie *Pulp Fiction*, a Quentin Tarantino flick that came out in 1994.

You must be partially awake for the surgery, so I decided to meditate. Someone was blasting that song in the operating room, and the giggling was rampant as the song looped. After sleeping it off, an older nurse came in wanting to shake my hand. She had never met a female longshore worker, before. I realized that laughter helped me to forget the fear of having my heart worked on.

Over the next three months I had several calls to prepare for the next step. The anesthesiologist appointment scared me because I was asked how much effort should be put into my resuscitation. That meant that I was going to be dead for a while. Yikes!

By May, the lobotomy called a COVID test was required three days prior to surgery, so it could be analyzed in time. My surgery got cancelled three times due to the priority of hearts becoming available for transplant. I got "lobotomized" three times in that parking lot, by the time it was my turn. I was about to turn fifty-seven, the following week.

CHAPTER 21

GETTING REBOOTED

I was guaranteed by my surgeon that he would personally call my husband to tell him how the open-heart surgery went. Under normal circumstances, your loved ones would be at the hospital, waiting anxiously. Having one anesthesiologist at the end of each arm was frightening. Once fear creeps in, it's meditation time. Plus, I hate needles. They hurt, and meditation helps me stay relaxed.

You are slowly put to sleep until your heart stops. For access to the heart, they saw your sternum down the center. Once your heart is fixed, they wire your sternum back together. It takes four and a half hours to perform this surgery. They called my husband when my heart started beating again.

After being in a coma for twenty-four hours in the intensive care ward, I was abruptly woken up by a kind individual who was brushing my teeth for me. Then, the washcloth that felt so good made me say, "Ooohhh! A whore's bath! Thank you so much!"

While the snickering was going on, I heard someone say that waking up with a sense of humor was a good sign.

I was put in a room with three others where I slept away another day. Monday was surgery, Tuesday was coma, Wednesday was bootcamp. It hadn't occurred to me that I had to learn how to move without using my chest or arms. I had gotten up and walked the hospital ward several times that day and I was tired out. I was deluding myself into thinking it would be easy.

Waking up on Thursday meant one more sleep until I was allowed to go home. It also meant that it was time to start lugging my ass up and down the stairs. That was the day I realized that I was sharing a room with comedians. We somehow started up about our inability to put underwear on. I mean, there was only one bathroom, and everyone's ass was hanging out. We made such a racket the nurses would come flying in demanding to know who was having a party on their heart ward. That just made us laugh more.

Before you get to go home, the stitches come out, which means that you must lie still for forty-five minutes, afterward. We were all warned to behave ourselves, beforehand. I must admit having fun like that made it hard to say goodbye. Once we got home, Zephyr wanted to smell my chest immediately. He became a hovercraft, watching me like a hawk all the time.

CHAPTER 22

AN ALIEN PRESENCE

Right away, I began dreaming of an alien bursting out of my chest. This dream started freaking me out, after having it a few times. I remember being afraid of falling back to sleep, once or twice.

I hated having to rely on my family for everything. Les had to cut up my food for frick's sake! I was required to be feeble for a whole month. I despise boredom so much that once I started doing housework again, I realized that I had missed how endless it can be.

The pandemic continued. A vaccine was in development by a few companies. My doctor told me that I should be good to return to work by November but any hopes of returning to normal disappeared along with Les's health. I was back at work, and Matt and I were doing our best taking turns looking after him. Sciatica can be debilitating.

It was one thing after another going into 2021, stupid health issues all the time. I had been yanking heavy steel doors open at work for several months before the pain in my chest began. Once it started, I felt it all the time. Eventually, moving my arms was next to impossible. Luckily, Les was feeling better by then.

When something goes wrong with your heart, you can spend days in emergency, getting test after test, until there is a diagnosis. A very strong antibiotic that takes four days for any effects to be felt, ended up fixing me. I was off work for another two months.

Not being allowed to interact socially took a toll on humanity, I think. Plus, all the grief of losing people. I was getting fed up with crying all the time. It had been several years since we had seen Miss and Raj. That was making me cry all the time, too. Being able to hear her face while seeing her voice became a running joke. It was nice to be able to do that, though.

When travel restrictions got lifted, Miss and Raj came up for Christmas. Remember how bad the weather was in December 2022? We got an extra four days because no one could fly anywhere.

As for me … in 2023, The Alien began emerging.

My name is Michele Kish, and I am an alien.

2036 THE ALIEN CHRONICLES: EPILOGUE

I wasn't going to stay in there while the planet suffered. Congratulations to all off-gridders! I have had visions of having to pounce on an elliptical machine to power up a toaster.

I think that losing the last of my aunts and uncles, plus a girlfriend this year, reminded me of how love and memories come with us and get left behind. Full circle.

Death before life and life after death. 2036. Tick tock.

The question seems to be, will I ever be able to go back to the purple planet? This is a very good question. It will take years to get enough power back for me to do that. I know one thing for sure, when I go back, I'm taking my physicality with me.

Until we meet again …

Printed in the USA
CPSIA information can be obtained
at www.ICGtesting.com
JSHW062147240324
59702JS00002B/16